Life in the United States

Contents

Chapter 1

Where Do You Live?

How can you describe where you live? Look at this map and find your state. Each region is separated into divisions. The map's key will show you how to find your **region** and your **division** of the United States.

United States Regions and Divisions

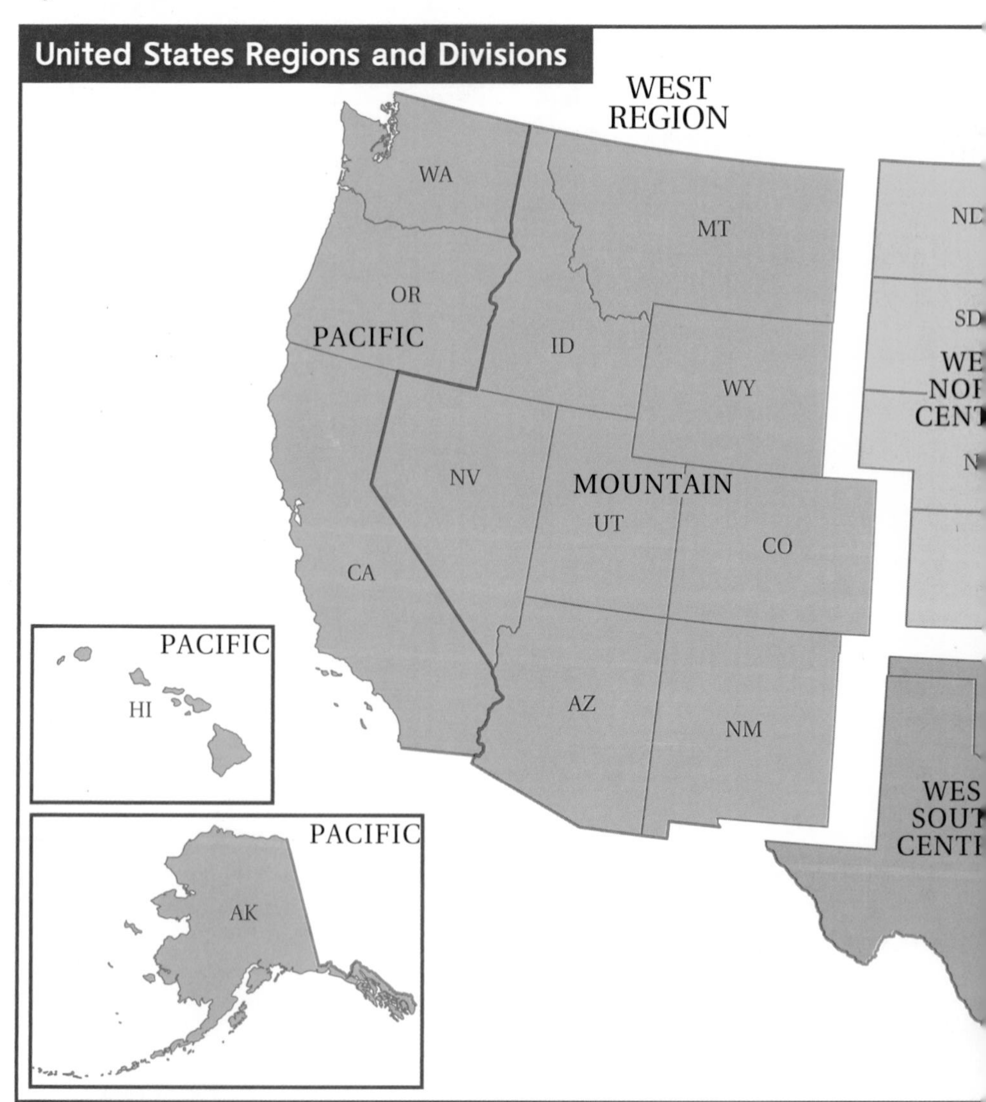

Each state is different. Some states are big. Few people live in them. Other states are small with many people living in them.

No matter how big or small, each state is special in some way.

Chapter 2

Life in the South

The South region has 16 states. More states are in this region than any other region. The South region is divided into three divisions: West South Central, East South Central, and South Atlantic.

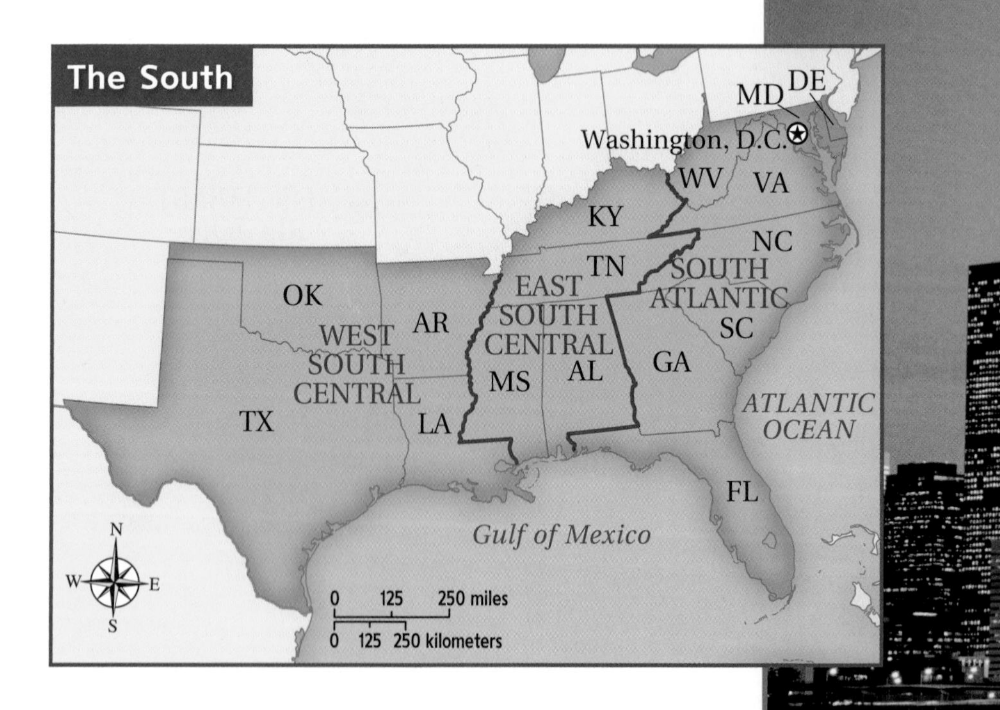

The states in the South region are different sizes. Some states are small, some are midsize, and one is big. The largest state in this region is Texas. It is the second largest state in the nation. Texas also has the second greatest population in the United States. The Texas city with the most people is Houston. Houston's population is about 2,010,000.

The second smallest South region state in size is Delaware. It is 96 miles long. Its width varies from 9 to 35 miles. Delaware has a small population.

Texas is a big state with a big population. Delaware is a small state with a small population.

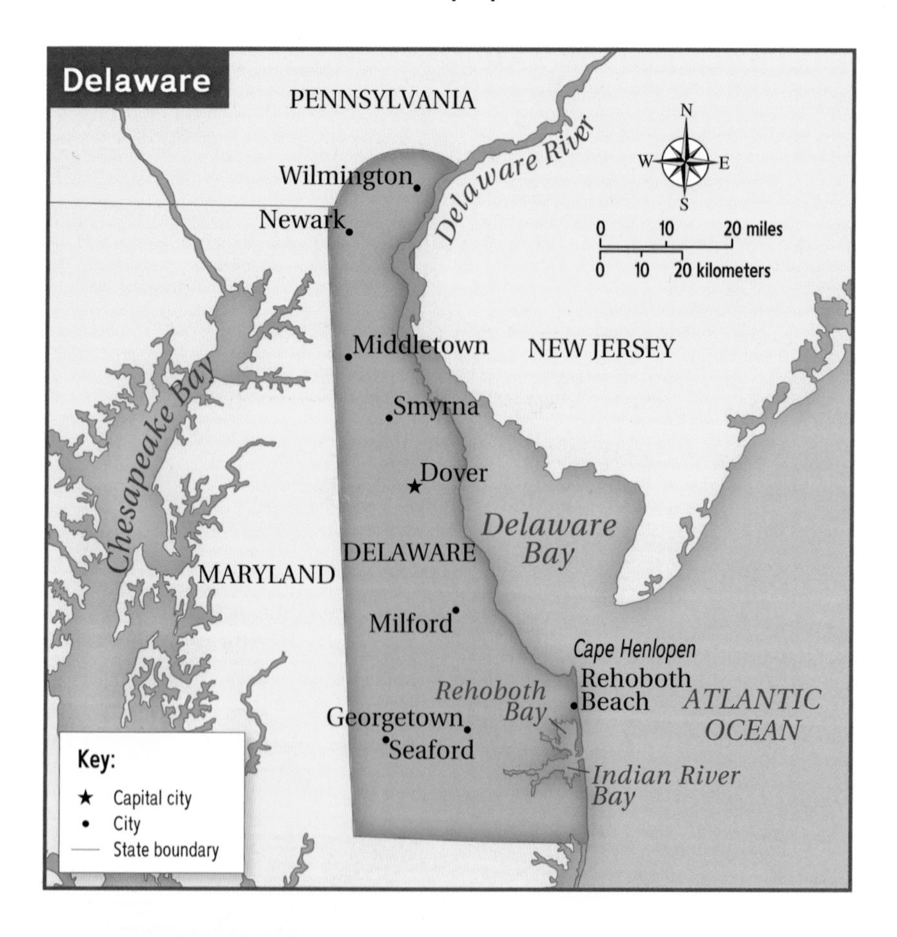

On December 7, 1787, Delaware was the first colony to ratify the United States Constitution.

Another state in the South region is Florida. Florida is medium in size. However, Florida has the 4th largest population in the United States.

Land Use and Resources

Delaware

- **Agriculture**: poultry, nursery stock, soybeans, dairy products, corn
- **Industry**: chemical products, food processing, paper products, rubber and plastic products, scientific instruments, printing and publishing

Florida

- **Agriculture**: citrus, vegetables, nursery stock, cattle, sugarcane, dairy products
- **Industry**: tourism, electric equipment, food processing, printing and publishing, transportation equipment, machinery

Texas

- **Agriculture**: cattle, cotton, dairy products, nursery stock, poultry, sorghum, corn, wheat
- **Industry**: chemical products, petroleum and natural gas, food processing, electric equipment, machinery, mining, tourism

Delaware, Florida, and Texas are different in many ways. The table shows what these 3 states have to offer.

Look at the chart. Which state has the greatest population per square mile, or **population density**? It is Delaware.

Delaware is much smaller in size than Florida and Texas. Fewer people live in Delaware. However, Delaware has more people per square mile.

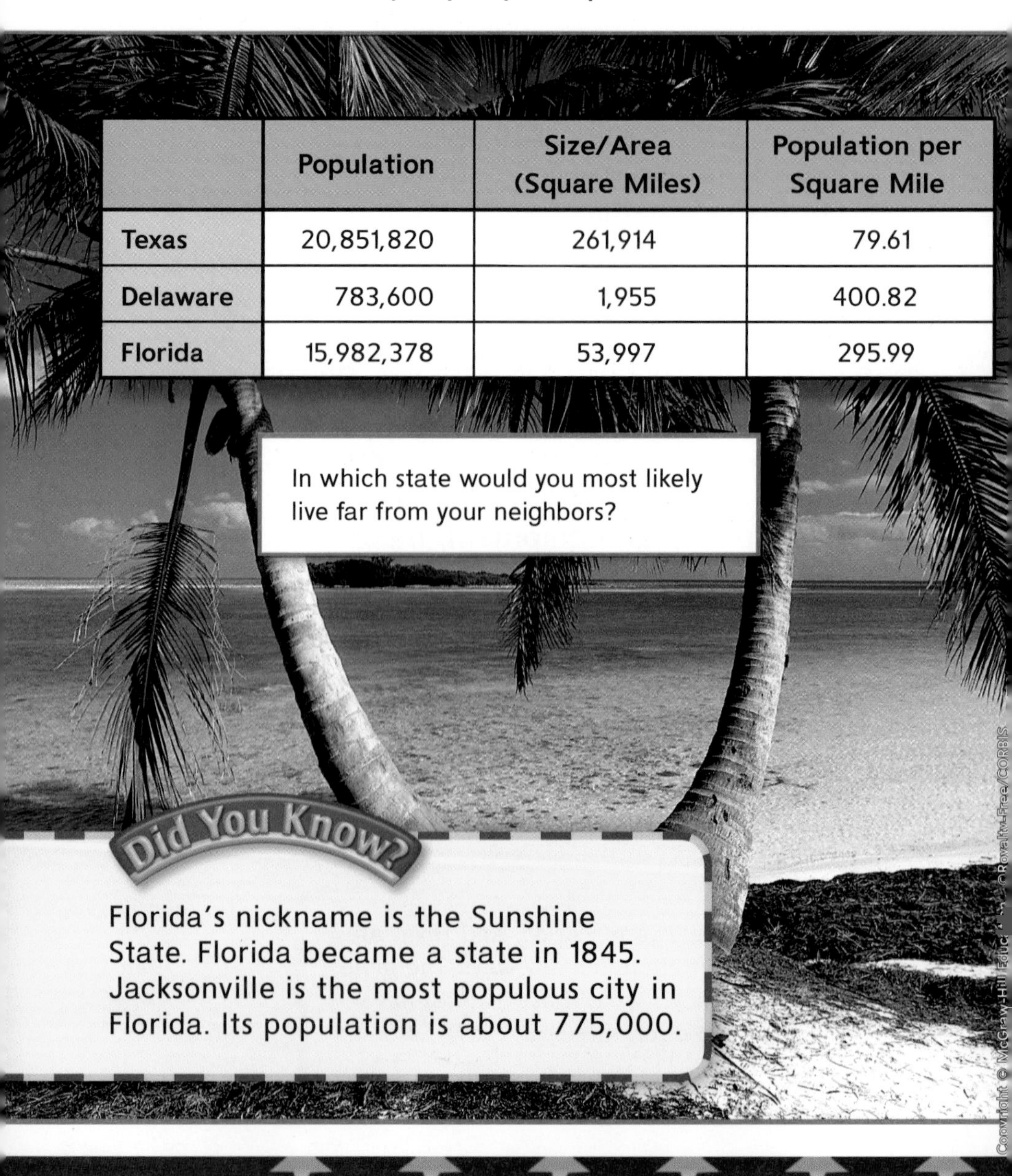

	Population	Size/Area (Square Miles)	Population per Square Mile
Texas	20,851,820	261,914	79.61
Delaware	783,600	1,955	400.82
Florida	15,982,378	53,997	295.99

In which state would you most likely live far from your neighbors?

Did You Know?

Florida's nickname is the Sunshine State. Florida became a state in 1845. Jacksonville is the most populous city in Florida. Its population is about 775,000.

Delaware, Florida, and Texas are all located along the United States coast. Fifty-three percent of the United States' population lives in **coastal** areas. This means that more than half, or 53 out of every 100, U.S. citizens live near the coast.

Many people like the coast. Coastal areas have many natural resources, diverse animal and plant species, job opportunities, recreation, tourism, trade routes, and industry.

Distribution of the Nation's Coastal Population

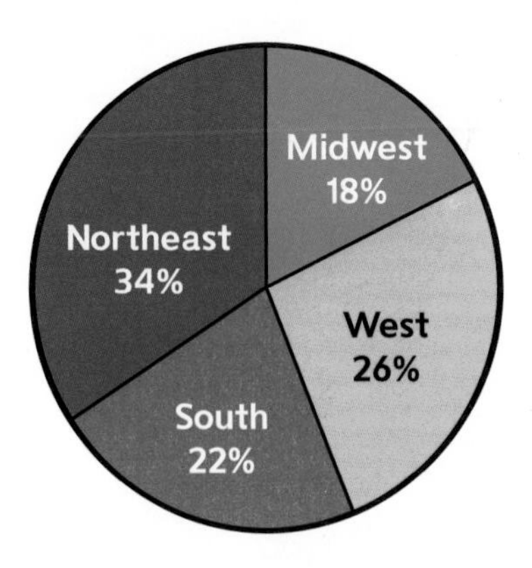

The circle graph shows which coastal areas are most popular to live in.

Ten of the 15 most populous cities in the United States are located in coastal areas.

Chapter 3

Life in the Midwest

The Midwest region is divided into two divisions: West North Central and East North Central. This region has 12 states.

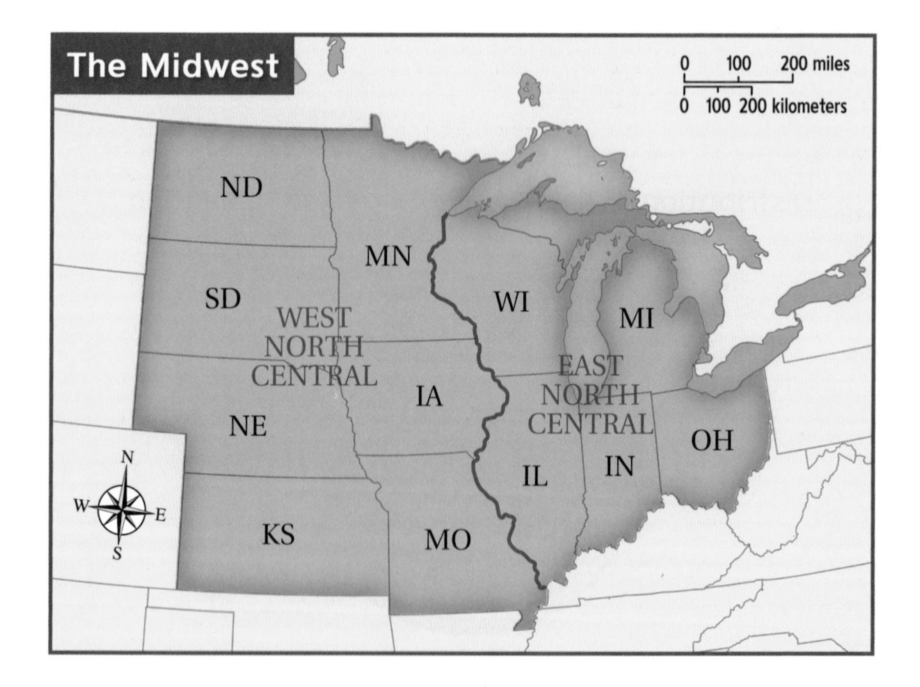

TALK ABOUT IT

Twelve of the 50 states of the nation are in the Midwest region. Write this figure as a fraction. Then write the fraction in simplest form.

Many Midwest states are similar in size. Some states are larger, some are a little smaller.

However, there is a big difference in the populations of these states. Of the Midwest states, Illinois has the greatest population. Chicago, the third largest city in the United States, is in Illinois. Kansas has the greatest area, with 81,815 total square miles.

Midwest State Nicknames	
Illinois	Prairie State
Indiana	Hoosier State
Iowa	Hawkeye State
Kansas	Sunflower State
Michigan	Wolverine State
Minnesota	North Star State
Missouri	Show Me State
Nebraska	Cornhusker State
North Dakota	Peace Garden State
Ohio	Buckeye State
South Dakota	Mount Rushmore State
Wisconsin	Badger State

Did You Know?

A buckeye is a nut that is found on a buckeye tree. Native Americans thought the markings on the nut look like the eye of a buck (a male deer), which gave this nut its name.

Five states in Midwest region are in the East North Central division. Seven states are in the West North Central division.

The land in this Midwest region is used mostly for farming and grazing.

Which division is larger in square miles? It is the West North Central division. However, the East North Central division has more than two times as many people.

Division	Population	Size/Area (Square Miles)	Population per Square Mile
West North Central	19,237,739	508,227	37.85
East North Central	45,155,037	243,539	185.41

TALK ABOUT IT

In which division are you more likely to rub elbows with other people?

How does the Midwest region's population compare to the South region's population? The total population of the South region is 100,236,820. The total population of the Midwest region is 64,392,776.

You can use estimation and a fraction to compare the populations.

$$\frac{\text{Midwest Region}}{\text{South Region}} \begin{matrix}\rightarrow \\ \rightarrow\end{matrix} \frac{64,392,776}{100,236,820} \rightarrow \frac{64,\cancel{000,000}}{100,\cancel{000,000}} = \frac{64}{100} = \frac{16}{25}$$

divide the numerator and denominator by 1,000,000

The map below can also help you compare the Midwest region to the South region. It shows the different population densities of the United States.

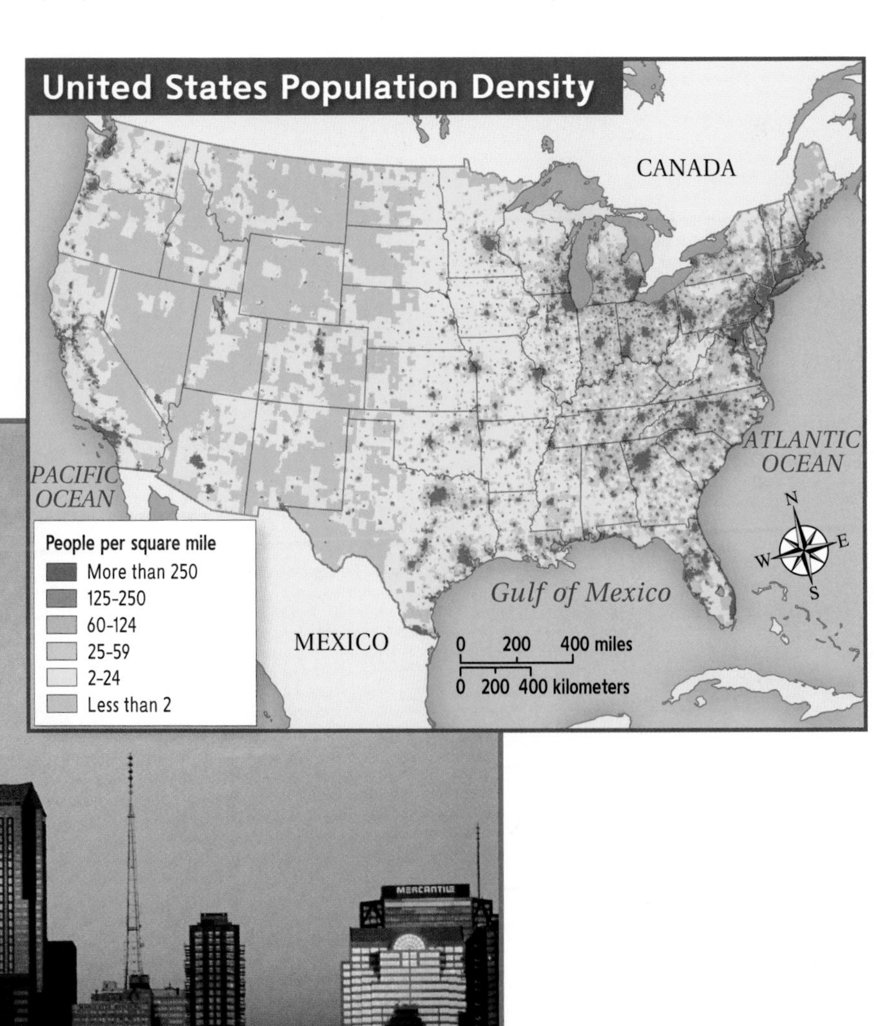

The Gateway Arch in St. Louis, MO

Life in the West

The West region has two divisions: Pacific and Mountain. This region has 13 states.

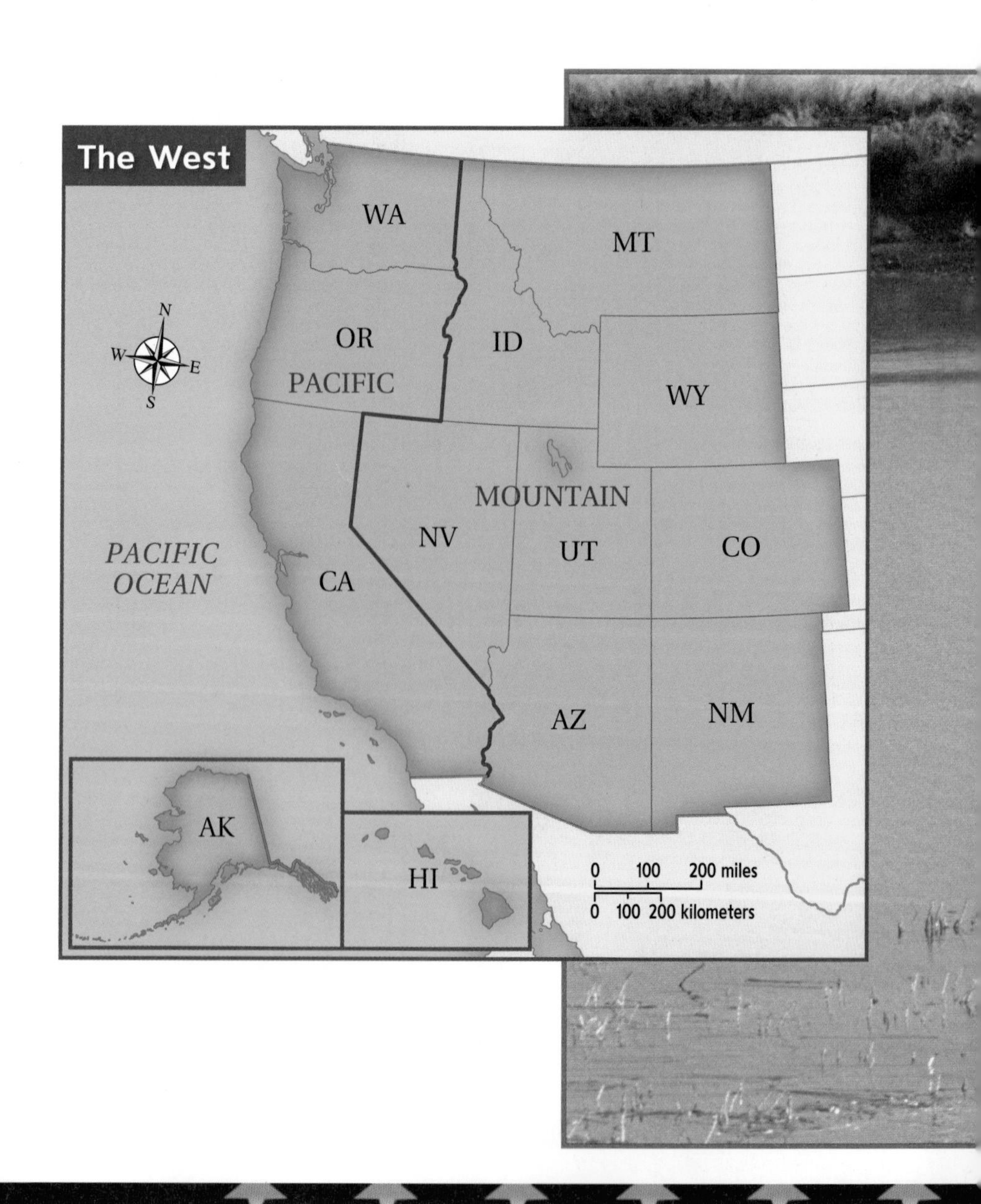

It may be hard to see on the map on page 16, but the state with the greatest area is in the West region. This state is Alaska.

Alaska has 570,374 square miles of land. That is more than two times the land area of Texas, which is the 2nd largest state. Alaska is about $\frac{1}{5}$ the size of all of the 48 **contiguous** (kon TIH gyoo us) states.

Bull moose in Alaska

Alaska ranks 48th in population. Only Vermont and Wyoming have fewer people. The United States purchased Alaska from Russia in 1867 for $7.2 million, or 2¢ per acre. This is a bargain when you know what Alaska has to offer.

Alaska Has a Lot to Offer

- Alaska's Tongass National Forest is the largest national forest in the U.S.
- Alaska accounts for 25% of the oil produced in the U.S.
- Salmon, crab, halibut, and herring come from Alaska.
- 17 of the 20 highest peaks in the U.S. are located in Alaska.
- Alaska has 15 national parks.

Alaska is the biggest state in size. California has the greatest population. This means California has more people living within its borders than any other state. Its population density is 12th out of the 50 states.

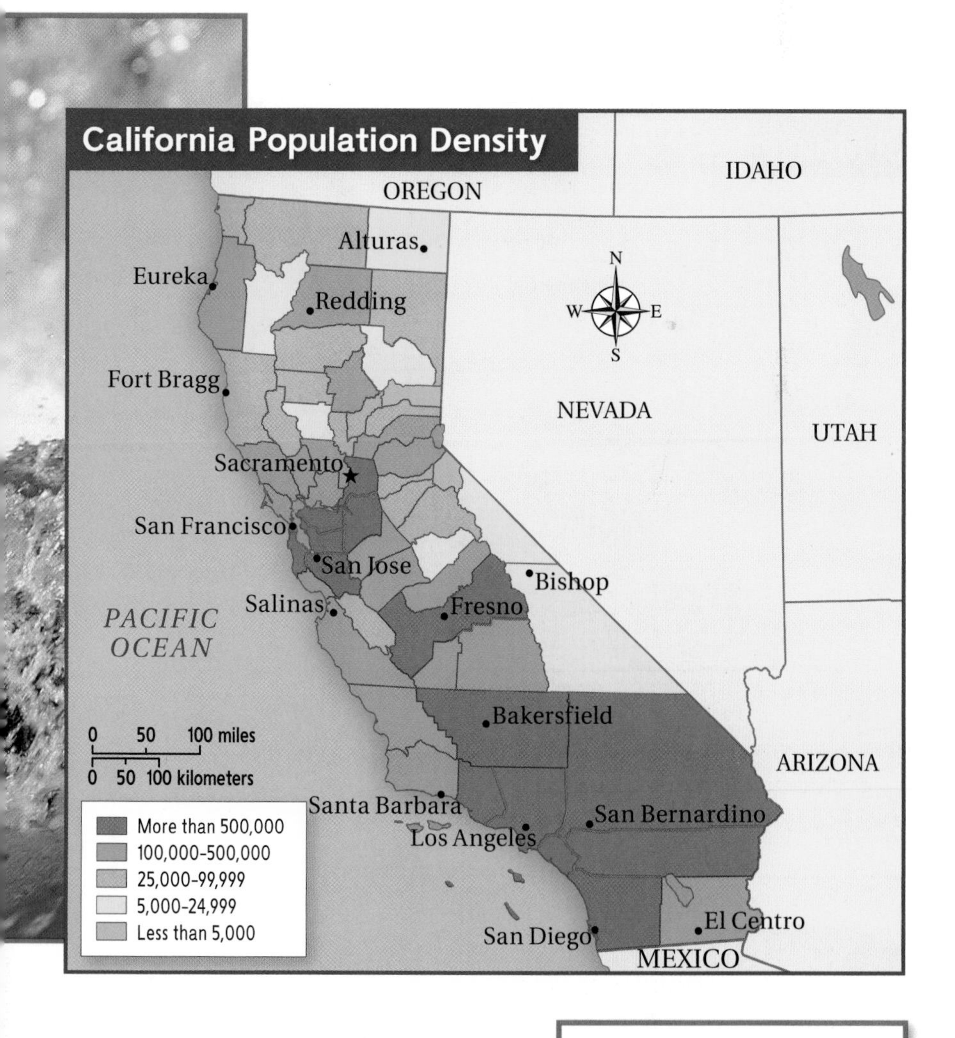

California's population density per square mile

Chapter 5

Life in the Northeast

The Northeast region is divided into two divisions: Middle Atlantic and New England. This region has 9 states. The Northeast region has the fewest states of all the regions.

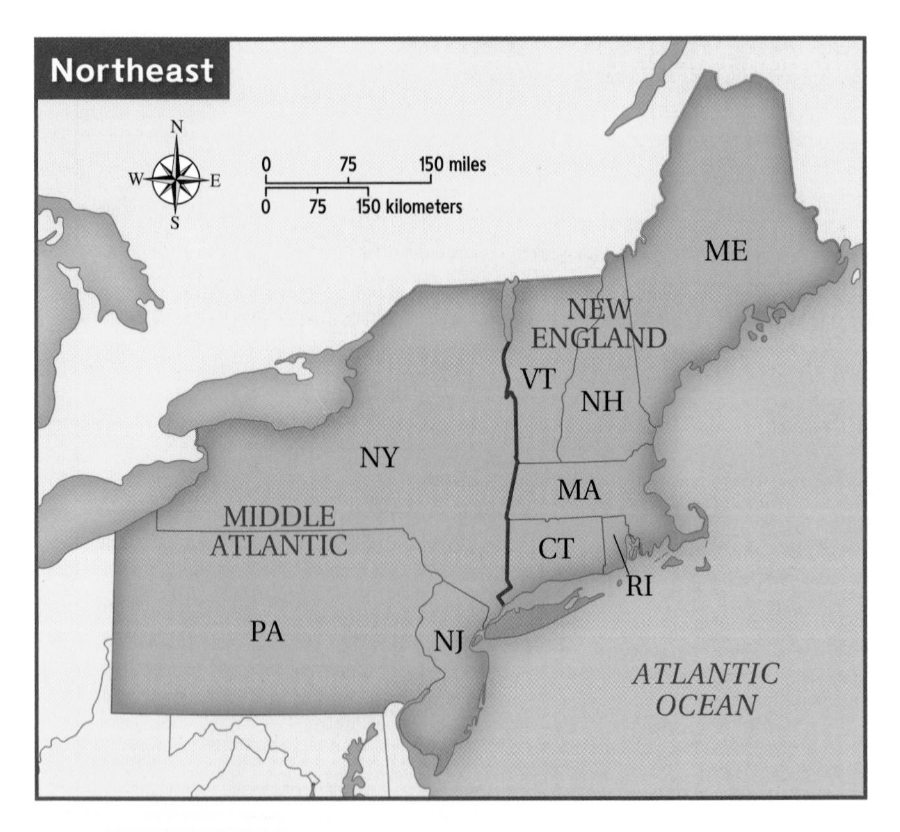

Rhode Island could fit into Alaska 425 times!

What do you notice about the size of the states in the Northeast region? Most of the states are small. Three Northeast states rank in the top 10 for population. New York is the 3rd most populous state in the nation. Pennsylvania is the 6th most populous state. New Jersey is 9th.

The Liberty Bell rang on July 8, 1776. It rang to summon the citizens of Philadelphia, Pennsylvania. They heard the first public reading of the Declaration of Independence.

New Jersey is the 9th most populated state. It is 46th on the list of states in order by area. How does New Jersey compare with the state of Alaska?

	Population	Size/Area (Square Miles)	Population per Square Mile
New Jersey	8,414,350	7,418	1134.32
Alaska	626,932	570,374	1.10

Alaska is a big state with a small population. New Jersey is a small state with a big population.

Atlantic City boardwalk in New Jersey

Chapter 6

Life in the United States

The following chart compares the four regions of the United States. The South region is the most populated. The West region has the most square miles. Which region has the most people per square mile? Which region has the least people per square mile?

State	Population	Size/Area (Square Miles)	Population per Square Mile
Northeast Region	53,594,378	162,272	330.28
West Region	63,197,932	1,751,477	36.08
Midwest Region	64,392,776	751,766	85.66
South Region	100,236,820	871,009	115.08

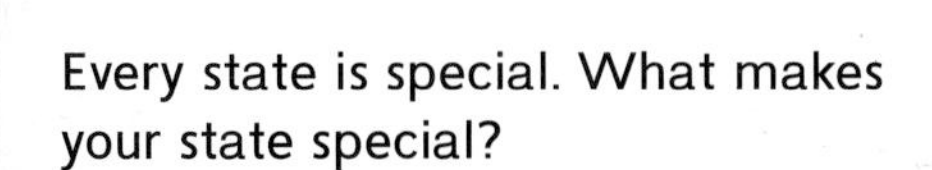
Every state is special. What makes your state special?

Glossary

coastal
Near a coast. *(page 9)*

contiguous
Touching or connected. *(page 17)*

division
One of the parts created when something is split. *(page 2)*

population density
The number of people per square mile. *(page 8)*

populous
Numerous in amount of people. *(page 5)*

region
An area with common features that set it apart from other areas. *(page 2)*